THE PRACTICE OF BUSINESS STATISTICS

COMPANION CHAPTER 17
LOGISTIC REGRESSION

David S. Moore
Purdue University

George P. McCabe
Purdue University

William M. Duckworth
Iowa State University

Stanley L. Sclove
University of Illinois

W. H. Freeman and Company
New York

Senior Acquisitions Editor:	**Patrick Farace**
Senior Developmental Editor:	**Terri Ward**
Associate Editor:	**Danielle Swearengin**
Media Editor:	**Brian Donnellan**
Marketing Manager:	**Jeffrey Rucker**
Head of Strategic Market Development:	**Clancy Marshall**
Project Editor:	**Mary Louise Byrd**
Cover and Text Design:	**Vicki Tomaselli**
Production Coordinator:	**Paul W. Rohloff**
Composition:	**Publication Services**
Manufacturing:	**RR Donnelley & Sons Company**

Cataloguing-in-Publication Data available from the Library of Congress

Library of Congress Control Number: 2002108463

Printed in the United States of America

Second Printing

TO THE INSTRUCTOR

NOW *YOU* HAVE THE CHOICE!

This is **Companion Chapter 17** to *The Practice of Business Statistics (PBS).* Please note that this chapter, along with any other Companion Chapters, can be bundled with the *PBS* Core book, which contains Chapters 1–11.

CORE BOOK

Chapter 1	Examining Distributions
Chapter 2	Examining Relationships
Chapter 3	Producing Data
Chapter 4	Probability and Sampling Distributions
Chapter 5	Probability Theory
Chapter 6	Introduction to Inference
Chapter 7	Inference for Distributions
Chapter 8	Inference for Proportions
Chapter 9	Inference for Two-Way Tables
Chapter 10	Inference for Regression
Chapter 11	Multiple Regression

These other **Companion Chapters,** *in any combinations you wish,* are available for you to package with the *PBS* Core book.

TAKE YOUR PICK

Chapter 12	Statistics for Quality: Control and Capability
Chapter 13	Time Series Forecasting
Chapter 14	One-Way Analysis of Variance
Chapter 15	Two-Way Analysis of Variance
Chapter 16	Nonparametric Tests
Chapter 17	Logistic Regression
Chapter 18	Bootstrap Methods and Permutation Tests

LOGISTIC REGRESSION

Prelude

Mail-order sales

A company uses many different mailed catalogs to sell different types of merchandise. One catalog that features home goods, such as bedspreads and pillows, was mailed to 200,000 people who were not current customers.[1] The response variable is whether or not the person places an order. Logistic regression is used to model the probability p of a purchase as a function of five explanatory variables. These are the number of purchases within the last 24 months from a home gift catalog, the proportion of single people in the zip code area based on census data, the number of credit cards, a variable that distinguishes apartment dwellers from those who live in single-family homes, and an indicator of whether or not the customer has ever made a purchase from a similar type of catalog. The fitted logistic model is then used to estimate the probability that a large collection of potential customers will make a purchase. Catalogs are sent to those whose estimated probability is above some cutoff value.

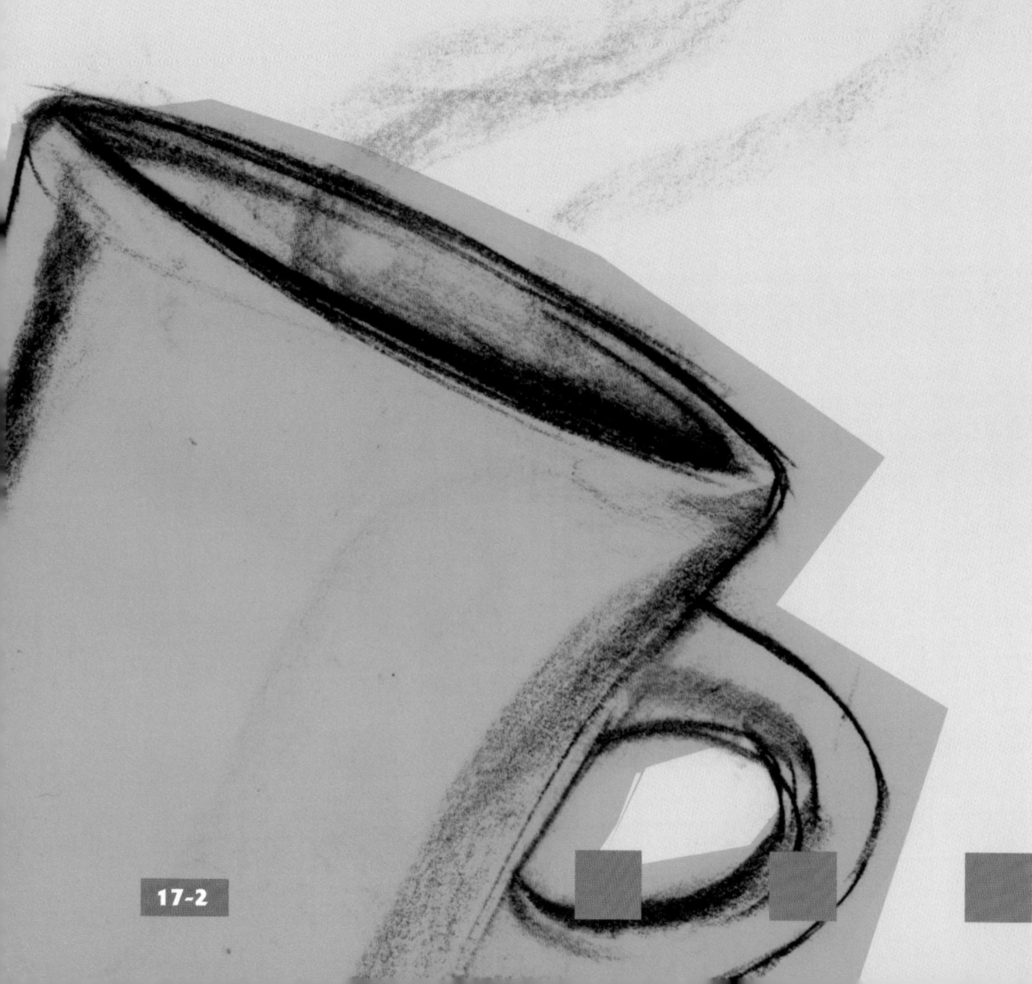

Logistic Regression*

*This chapter requires the material on binomial distributions in the optional Chapter 5.

Introduction

The simple and multiple linear regression methods we studied in Chapters 10 and 11 are used to model the relationship between a quantitative response variable and one or more explanatory variables. In this chapter we describe similar methods for use when the response variable has only two possible values: customer buys or does not buy, patient lives or dies, candidate accepts job or not.

In general, we call the two values of the response variable "success" and "failure" and represent them by 1 (for a success) and 0. The mean is then the proportion of ones, $p = P(\text{success})$. If our data are n independent observations with the same p, this is the *Binomial setting* (page 319). What is new in this chapter is that the data now include an *explanatory variable* x. The probability p of a success depends on the value of x. For example, suppose we are studying whether a customer makes a purchase ($y = 1$) or not ($y = 0$) after being offered a discount. Then p is the probability that the customer makes a purchase, and possible explanatory variables include (a) whether the customer has made similar purchases in the past, (b) the type of discount, and (c) the age of the customer. Note that the explanatory variables can be either categorical or quantitative. Logistic regression[2] is a statistical method for describing these kinds of relationships.

17.1 The Logistic Regression Model

In general, the data for logistic regression are n independent observations, each consisting of a value of the explanatory variable x and either a success or a failure for that trial. For example, x may be the age of a customer, and "success" means that this customer made a purchase. Every observation may have a different value of x. To introduce logistic regression, however, it is convenient to start with the special case in which the explanatory variable x is also a yes-or-no variable. The data then contain a number of outcomes (success or failure) for each of the two values of x. There are just two values of p, one for each value of x. The count of successes for each value of x has a Binomial distribution, so that we are on familiar ground. Here is an example.

CASE 17.1

BINGE DRINKERS

Exercise 8.14 (page 517) describes a survey of 17,096 students in U.S. four-year colleges. A student who reports drinking five or more drinks in a row three or more times in the past two weeks is called a "frequent binge drinker." The researchers were interested in estimating the proportion of students who are frequent binge drinkers.

indicator variable

One promising explanatory variable is the gender of the student. We express gender numerically using an **indicator variable,**

$$x = \begin{cases} 1 & \text{if the student is a man} \\ 0 & \text{if the student is a woman} \end{cases}$$

The sample contained 7180 men and 9916 women. The probability that a randomly chosen student is a frequent binge drinker has two values, p_1 for men and p_0 for women. The number of men in the sample who are frequent binge drinkers has the Binomial distribution $B(7180, p_1)$. The count of frequent binge drinkers among the women has the $B(9916, p_0)$ distribution.

Binomial distributions and odds

We begin with a review of some ideas associated with Binomial distributions.

EXAMPLE 17.1 **Proportion of binge drinkers**

CASE 17.1

In Chapter 8 we used sample proportions to estimate population proportions. The binge-drinking study found that 1630 of the 7180 men in the sample were frequent binge drinkers, as were 1684 of the 9916 women. Our estimates of the two population proportions are

$$\text{Men: } \hat{p}_1 = \frac{1630}{7180} = 0.2270$$

and

$$\text{Women: } \hat{p}_0 = \frac{1684}{9916} = 0.1698$$

That is, we estimate that 22.7% of college men and 17.0% of college women are frequent binge drinkers.

odds

Logistic regression works with **odds** rather than proportions. The odds are the ratio of the proportions for the two possible outcomes. If p is the probability of a success, then $1 - p$ is the probability of a failure, and

$$\text{ODDS} = \frac{p}{1-p} = \frac{\text{probability of success}}{\text{probability of failure}}$$

A similar formula for the sample odds is obtained by substituting $\hat{p}$ for p in this expression.

EXAMPLE 17.2 **Odds of being a binge drinker**

CASE 17.1

The proportion of frequent binge drinkers among the men in the study is $\hat{p}_1 = 0.227$, so the proportion of men who are not frequent binge drinkers is

$$1 - \hat{p}_1 = 1 - 0.2270 = 0.7730$$

The estimated odds of a male student being a frequent binge drinker are therefore

$$\text{ODDS} = \frac{\hat{p}_1}{1 - \hat{p}_1}$$

$$= \frac{0.2270}{0.7730} = 0.2937$$

For women, the odds are

$$\text{ODDS} = \frac{\hat{p}_0}{1 - \hat{p}_0}$$

$$= \frac{0.1698}{1 - 0.1698} = 0.2045$$

When people speak about odds, they often round to integers or fractions. Since 0.205 is approximately 1/5, we could say that the odds that a female college student is a frequent binge drinker are 1 to 5. In a similar way, we could describe the odds that a college woman is *not* a frequent binge drinker as 5 to 1.

17.1 Successful franchises and exclusive territories. In Case 9.1 (page 549) we studied data on the success of 170 franchise firms and whether or not the owner of a franchise had an exclusive territory. Here are the data:

Observed numbers of firms

	Exclusive territory		
Success	Yes	No	Total
Yes	108	15	123
No	34	13	47
Total	142	28	170

What proportion of the exclusive-territory firms are successful? Find the proportion for the firms that do not have exclusive territories. Convert each of these proportions to odds.

17.2 No Sweat labels and gender. In Example 9.2 (page 548) we examined the relationship between gender and the use of clothing labels that indicated that the garments were manufactured under fair working conditions. Here are the data:

	Gender		
Label user	Women	Men	Total
Yes	63	27	90
No	233	224	457
Total	296	251	547

Find the proportion of men who are label users; do the same for women. Restate each of these proportions as odds.

Model for logistic regression

In Chapter 8 we learned how to compare the proportions of frequent binge drinkers among college men and women using z tests and confidence intervals. Logistic regression is another way to make this comparison, one that extends to more general settings with a success-or-failure response variable.

In simple linear regression we modeled the mean μ of the response variable y as a linear function of the explanatory variable: $\mu = \beta_0 + \beta_1 x$. When y is just 1 or 0 (success or failure), the mean is the probability p of a success. Logistic regression models the mean p in terms of an explanatory variable x. We might try to relate p and x as in simple linear regression: $p = \beta_0 + \beta_1 x$. Unfortunately, this is not a good model. Whenever $\beta_1 \neq 0$, extreme values of x will give values of $\beta_0 + \beta_1 x$ that fall outside the range of possible values of p, $0 \leq p \leq 1$.

log odds

The logistic regression model removes this difficulty by working with the natural logarithm of the odds, $p/(1 - p)$. We use the term **log odds** for this transformation. As p moves from 0 to 1, the log odds moves through all negative and positive numerical values. We model the log odds as a linear function of the explanatory variable:

$$\log\left(\frac{p}{1-p}\right) = \beta_0 + \beta_1 x$$

Figure 17.1 graphs the relationship between p and x for some different values of β_0 and β_1. The logistic regression model uses *natural* logarithms. Most calculators and statistical software systems have a built-in function for the natural logarithm, often labeled "ln."

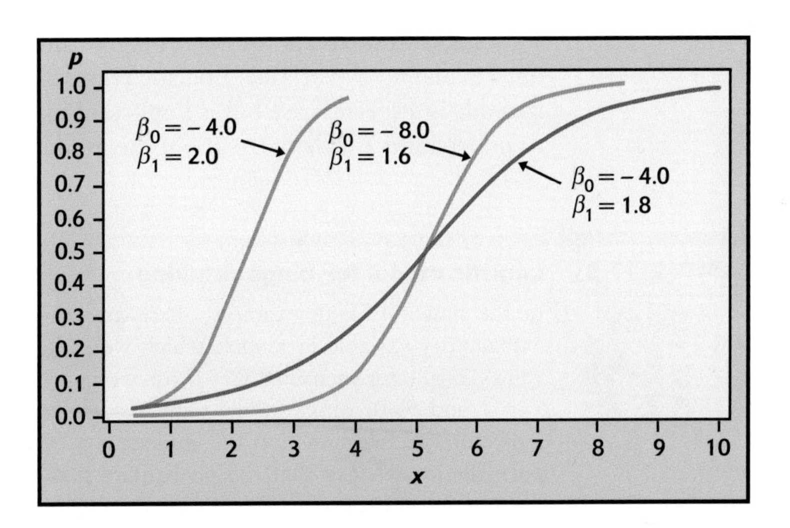

FIGURE 17.1 Plot of p versus x for selected values of β_0 and β_1.

Returning to the binge-drinking study, we have for men

$$\log(\text{ODDS}) = \log(0.2937) = -1.23$$

and for women

$$\log(\text{ODDS}) = \log(0.2045) = -1.59$$

Verify these results with your calculator, remembering that "log" is the natural logarithm. Here is a summary of the logistic regression model.

LOGISTIC REGRESSION MODEL

The **statistical model for logistic regression** is

$$\log\left(\frac{p}{1-p}\right) = \beta_0 + \beta_1 x$$

where p is a Binomial proportion and x is the explanatory variable. The parameters of the logistic model are β_0 and β_1.

APPLY YOUR KNOWLEDGE

17.3 **Log odds for exclusive territories.** Refer to Exercise 17.1. Find the log odds for the franchises that have exclusive territories. Do the same for the firms that do not.

17.4 **Log odds for No Sweat labels.** Refer to Exercise 17.2. Find the log odds for the women. Do the same for the men.

Fitting and interpreting the logistic regression model

We must now fit the logistic regression model to data. In general, the data consist of n observations on the explanatory variable x, each with a success-or-failure response. Our binge-drinking example has an indicator (0 or 1) explanatory variable. Logistic regression with an indicator explanatory variable is a special case but is important in practice. We use this special case to understand a little more about the model.

EXAMPLE 17.3 **Logistic model for binge drinking**

CASE 17.1

In the binge-drinking example, there are $n = 17{,}096$ observations. The explanatory variable is gender, which we coded using an indicator variable with values $x = 1$ for men and $x = 0$ for women. There are 7180 observations with $x = 1$ and 9916 observations with $x = 0$. The response variable is also an indicator variable: $y = 1$ if the student is a frequent binge drinker and $y = 0$ if not. The model says that the probability p that a student is a frequent binge drinker depends on the student's gender ($x = 1$ or $x = 0$). There are two possible values for p, say p_1 for men and p_0 for women. The model says that for men

$$\log\left(\frac{p_1}{1-p_1}\right) = \beta_0 + \beta_1$$

and for women

$$\log\left(\frac{p_0}{1-p_0}\right) = \beta_0$$

Note that there is a β_1 term in the equation for men because $x = 1$, but it is missing in the equation for women because $x = 0$.

In general, the calculations needed to find estimates b_0 and b_1 for the parameters β_0 and β_1 are complex and require the use of software. When the explanatory variable has only two possible values, however, we can easily find the estimates. This simple framework also provides a setting where we can learn what the logistic regression parameters mean.

EXAMPLE 17.4 **Parameter estimates for binge drinking**

For the binge-drinking example, we found the log odds for men,

$$\log\left(\frac{\hat{p}_1}{1-\hat{p}_1}\right) = -1.22$$

and for women,

$$\log\left(\frac{\hat{p}_0}{1-\hat{p}_0}\right) = -1.58$$

To find estimates b_0 and b_1 of the model parameters β_0 and β_1, we match the male and female model equations in Example 17.3 with the corresponding data equations. Because

$$\log\left(\frac{p_0}{1-p_0}\right) = \beta_0 \quad \text{and} \quad \log\left(\frac{\hat{p}_0}{1-\hat{p}_0}\right) = -1.58$$

the estimate b_0 of the intercept is simply the log(ODDS) for the women,

$$b_0 = -1.58$$

Similarly, the estimated slope is the difference between the log(ODDS) for the men and the log(ODDS) for the women,

$$b_1 = -1.22 - (-1.58) = 0.36$$

The fitted logistic regression model is

$$\log(\text{ODDS}) = -1.58 + 0.36x$$

The slope in this logistic regression model is the difference between the log(ODDS) for men and the log(ODDS) for women. Most people are not comfortable thinking in the log(ODDS) scale, so interpretation of the results in terms of the regression slope is difficult. Usually, we apply a transformation to help us. The exponential function (e^x key on your calculator) reverses the natural logarithm transformation. That is, continuing Example 17.4,

$$\text{ODDS} = e^{-1.58+0.36x} = e^{-1.58} \times e^{0.36x}$$

From this, the ratio of the odds for men ($x = 1$) and women ($x = 0$) is

$$\frac{\text{ODDS}_{\text{men}}}{\text{ODDS}_{\text{women}}} = e^{0.36} = 1.43$$

odds ratio

The transformation $e^{0.36}$ undoes the logarithm and transforms the logistic regression slope into an **odds ratio,** in this case, the ratio of the odds that a man is a frequent binge drinker to the odds that a woman is a frequent binge drinker. We can multiply the odds for women by the odds ratio to obtain the odds for men:

$$\text{ODDS}_{\text{men}} = 1.43 \times \text{ODDS}_{\text{women}}$$

The odds for men are 1.43 times the odds for women.

Notice that we have chosen the coding for the indicator variable so that the regression slope is positive. This will give an odds ratio that is greater than 1. Had we coded women as 1 and men as 0, the signs of the parameters would be reversed, the fitted equation would be $\log(\text{ODDS}) = 1.58 - 0.36x$, and the odds ratio would be $e^{-0.36} = 0.70$. The odds for women are 70% of the odds for men.

It is of course often the case that the explanatory variable is quantitative rather than an indicator variable. We must then use software to fit the logistic regression model. Here is an example.

EXAMPLE 17.5 **Is the quality of the product acceptable?**

The CHEESE data set described in the Data Appendix includes a response variable called "Taste" that is a measure of the quality of the cheese in the opinions of several tasters. For this example, we will classify the cheese as acceptable ($y = 1$) if Taste ≥ 37 and unacceptable ($y = 0$) if Taste < 37. This is our response variable. The data set contains three explanatory variables: "Acetic," "H2S," and "Lactic." Let's use Acetic as the explanatory variable. The model is

$$\log\left(\frac{p}{1-p}\right) = \beta_0 + \beta_1 x$$

where p is the probability that the cheese is acceptable and x is the value of Acetic. The model for estimated log odds fitted by software is

$$\log(\text{ODDS}) = b_0 + b_1 x = -13.71 + 2.25x$$

The odds ratio is $e^{b_1} = 9.49$. This means that if we increase the acetic acid content x by one unit, we increase the odds that the cheese will be acceptable by about 9.5 times. (See Exercise 17.7.)

APPLY YOUR KNOWLEDGE

17.5 **Fitted model for exclusive territories.** Refer to Exercises 17.1 and 17.3. Find the estimates b_0 and b_1 and give the fitted logistic model. What is the odds ratio for exclusive territory ($x = 1$) versus no exclusive territory ($x = 0$)?

17.6 **Fitted model for No Sweat labels.** Refer to Exercises 17.2 and 17.4. Find the estimates b_0 and b_1 and give the fitted logistic model. What is the odds ratio for women ($x = 1$) versus men ($x = 0$)?

17.7 **Interpreting an odds ratio.** If we apply the exponential function to the fitted model in Example 17.5, we get

$$\text{ODDS} = e^{-13.71+2.25x} = e^{-13.71} \times e^{2.25x}$$

Show that for any value of the quantitative explanatory variable x, the odds ratio for increasing x by 1,

$$\frac{\text{ODDS}_{x+1}}{\text{ODDS}_x}$$

is $e^{2.25} = 9.49$. This justifies the interpretation given at the end of Example 17.5.

Section 17.1 Summary

■ **Logistic regression** explains a success-or-failure response variable in terms of an explanatory variable.

■ If p is a proportion of successes, then the **odds** of a success are $p/(1 - p)$, the ratio of the proportion of successes to the proportion of failures.

■ The **logistic regression model** relates the proportion of successes in the population to an explanatory variable x through the logarithm of the odds of a success:

$$\log\left(\frac{p}{1-p}\right) = \beta_0 + \beta_1 x$$

That is, each value of x gives a different proportion p of successes. The **data** are n values of x, with observed success or failure for each. The model assumes that these n success-or-failure trials are independent, with probabilities of success given by the logistic regression equation. The **parameters** of the model are β_0 and β_1.

■ The **odds ratio** is the ratio of the odds of a success at $x + 1$ to the odds of a success at x. It is found as e^{β_1}, where β_1 is the slope in the logistic regression equation.

■ Software fits the data to the model, producing estimates b_0 and b_1 of the parameters β_0 and β_1.

17.2 Inference for Logistic Regression

Statistical inference for logistic regression with one explanatory variable is similar to statistical inference for simple linear regression. We calculate estimates of the model parameters and standard errors for these estimates. Confidence intervals are formed in the usual way, but we use standard Normal z^*-values rather than critical values from the t distributions. The ratio of the estimate to the standard error is the basis for hypothesis tests. Software often reports the test statistics as the squares of these ratios, in which case the P-values are obtained from the chi-square distribution with 1 degree of freedom.

CONFIDENCE INTERVALS AND SIGNIFICANCE TESTS FOR LOGISTIC REGRESSION

An approximate **level C confidence interval for the slope β_1** in the logistic regression model is

$$b_1 \pm z^* \mathrm{SE}_{b_1}$$

The ratio of the odds for a value of the explanatory variable equal to $x + 1$ to the odds for a value of the explanatory variable equal to x is the **odds ratio e^{β_1}**. A **level C confidence interval for the odds ratio** is obtained by transforming the confidence interval for the slope,

$$(e^{b_1 - z^* \mathrm{SE}_{b_1}},\ e^{b_1 + z^* \mathrm{SE}_{b_1}})$$

In these expressions z^* is the standard Normal critical value with area C between $-z^*$ and z^*.

To test the hypothesis H_0: $\beta_1 = 0$, compute the **test statistic**

$$X^2 = \left(\frac{b_1}{\mathrm{SE}_{b_1}}\right)^2$$

In terms of a random variable χ^2 having the χ^2 distribution with 1 degree of freedom, the P-value for a test of H_0 against H_a: $\beta_1 \neq 0$ is approximately $P(\chi^2 \geq X^2)$.

We have expressed the null hypothesis in terms of the slope β_1 because this form closely resembles the hypothesis of "no linear relationship" in simple linear regression. In many applications, however, the results are expressed in terms of the odds ratio. A slope of 0 is the same as an odds ratio of 1, so we often express the null hypothesis of interest as "the odds ratio is 1." This means that the two odds are equal and the explanatory variable is not useful for predicting the odds.

EXAMPLE 17.6 **Computer output for binge drinking**

CASE 17.1

Figure 17.2 gives the output from SPSS and SAS for the binge-drinking study. The parameter estimates given by SPSS are $b_0 = -1.587$ and $b_1 = 0.362$, more exact than we calculated directly in Example 17.4. The standard errors are 0.027 and 0.039. A 95% confidence interval for the slope is

$$b_1 \pm z^* \mathrm{SE}_{b_1} = 0.362 \pm (1.96)(0.039)$$
$$= 0.362 \pm 0.076$$

We are 95% confident that the slope is between 0.286 and 0.438. The SPSS output provides the odds ratio 1.435 (under the heading "Exp(B)") but does not give the confidence interval for e^{β_1}. This is easy to compute from the interval for the slope β_1:

$$(e^{b_1 - z^* \mathrm{SE}_{b_1}},\ e^{b_1 + z^* \mathrm{SE}_{b_1}}) = (e^{0.286},\ e^{0.438})$$
$$= (1.33,\ 1.55)$$

We conclude, "College men are more likely to be frequent binge drinkers than college women (odds ratio = 1.44, 95% CI = 1.33 to 1.55)."

It is standard to use 95% confidence intervals, and software often reports these intervals. A 95% confidence interval for the odds ratio also provides a

SPSS

Variables in the Equation

		B	S.E.	Wald	df	Sig.	Exp(B)
Step 1	GENDERM	0.362	0.039	86.611	1	0.000	1.435
	Constant	−1.587	0.027	3520.069	1	0.000	0.205

a Variable(s) entered on step 1: GENDERM

SAS

The LOGISTIC Procedure

Analysis of Maximum Likelihood Estimates

Parameter	DF	Estimate	Standard Error	Wald Chi-Square	Pr > ChiSq
Intercept	1	−1.5868	0.0267	3520.3120	< 0.001
genderm	1	0.3617	0.0388	86.6811	< 0.001

Odds Ratio Estimates

Effect	Point Estimate	95% Wald Confidence Limits	
genderm	1.436	1.330	1.549

FIGURE 17.2 Logistic regression output from SPSS and SAS for the binge-drinking data, for Example 17.6.

test of the null hypothesis that the odds ratio is 1 at the 5% significance level. If the confidence interval does not include 1, we reject H_0 and conclude that the odds for the two groups are different; if the interval does include 1, the data do not provide enough evidence to distinguish the groups in this way.

APPLY YOUR KNOWLEDGE

17.8 **Read the output.** Examine the SAS output in Figure 17.2. Report the estimates of β_0 and β_1 with the standard errors as given in this display. Also report the odds ratio with its 95% confidence interval as given in this output.

17.9 **Inference for exclusive territories.** Use software to run a logistic regression analysis for the exclusive territory data of Exercise 17.1. Summarize the results of the inference.

17.10 **Inference for No Sweat labels.** Use software to run the logistic regression analysis for the No Sweat label data of Exercise 17.2. Summarize the results of the inference.

Examples of logistic regression analyses

The following example is typical of many applications of logistic regression. It concerns a designed experiment with five different values for the explanatory variable.

EXAMPLE 17.7 **Effectiveness of an insecticide**

An experiment was designed to examine how well the insecticide rotenone kills an aphid, called *Macrosiphoniella sanborni*, that feeds on the chrysanthemum plant.[3] The explanatory variable is the concentration (in log of milligrams per liter) of the insecticide. About 50 aphids were exposed to each of five concentrations. Each insect was either killed or not killed. Although logistic regression uses the log odds of the proportion killed, software requires that we enter just the concentration, the number of insects, and the number killed. Here are the data, along with the results of some calculations:

Concentration x (log scale)	Number of insects	Number killed	Proportion killed $\hat{p}$	log odds
0.96	50	6	0.1200	−1.9924
1.33	48	16	0.3333	−0.6931
1.63	46	24	0.5217	0.0870
2.04	49	42	0.8571	1.7918
2.32	50	44	0.8800	1.9924

Least-squares regression of log odds on log concentration gives the fit illustrated in Figure 17.3. There is a clear linear relationship. The logistic regression fit for the proportion killed appears in Figure 17.4. It is a transformed version of Figure 17.3 with the fit calculated using the logistic model rather than by least-squares.

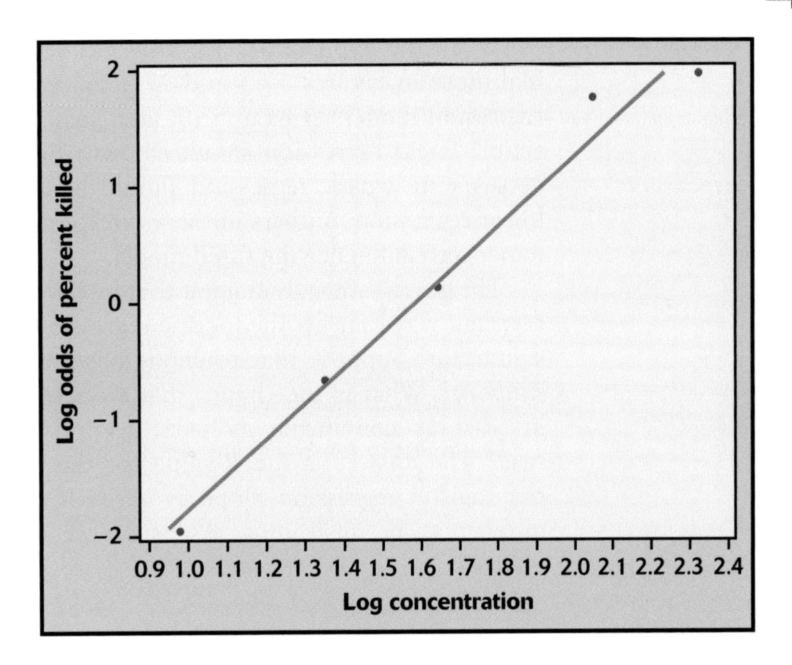

FIGURE 17.3 Plot of log odds of percent killed versus log concentration for the insecticide data, for Example 17.7.

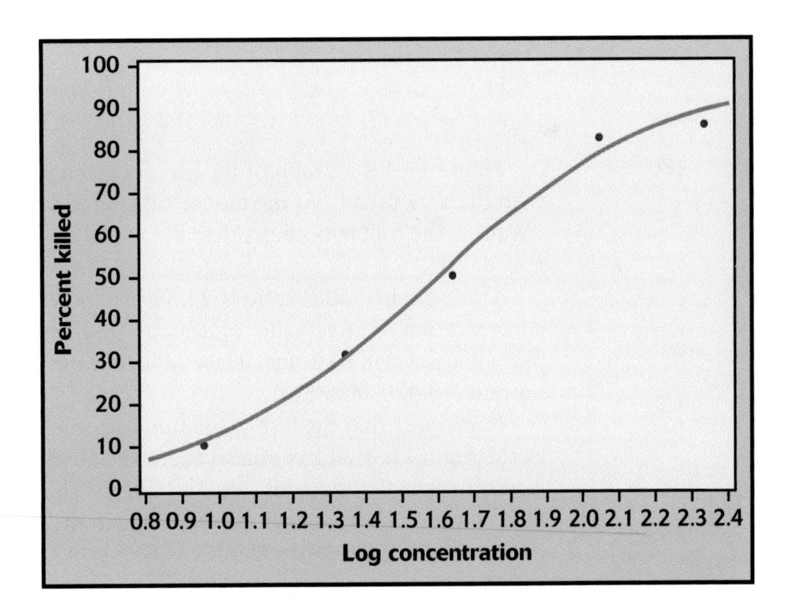

FIGURE 17.4 Plot of the percent killed versus log concentration with the logistic fit for the insecticide data, for Example 17.7.

When the explanatory variable has several values, we can use graphs like those in Figures 17.3 and 17.4 to visually assess whether the logistic regression model seems appropriate. Just as a scatterplot of y versus x in simple linear regression should show a linear pattern, a plot of log odds versus x in logistic regression should be close to linear. Just as in simple linear regression, outliers in the x direction should be avoided because they may overly influence the fitted model.

The graphs strongly suggest that insecticide concentration affects the kill rate in a way that fits the logistic regression model. Is the effect statistically significant? Suppose that rotenone has no ability to kill *Macrosiphoniella sanborni*. What is the chance that we would observe experimental results at least as convincing as what we observed if this supposition were true? The answer is the P-value for the test of the null hypothesis that the logistic regression slope is zero. If this P-value is not small, our graph may be misleading. As usual, we must add inference to our data analysis.

EXAMPLE 17.8 Does concentration affect the kill rate?

Figure 17.5 gives the output from SPSS, SAS, and Minitab for logistic regression analysis of the insecticide data. The model is

$$\log\left(\frac{p}{1-p}\right) = \beta_0 + \beta_1 x$$

where the values of the explanatory variable x are 0.96, 1.33, 1.63, 2.04, 2.32. From the SPSS output we see that the fitted model is

$$\log(\text{ODDS}) = b_0 + b_1 x = -4.89 + 3.11x$$

or

$$\frac{\hat{p}}{1-\hat{p}} = e^{-4.89+3.11x}$$

Figure 17.4 is a graph of the fitted $\hat{p}$ given by this equation against x, along with the data used to fit the model. SPSS gives the statistic X^2 under the heading "Wald." The null hypothesis that $\beta_1 = 0$ is clearly rejected ($X^2 = 64.23$, $P < 0.001$).

The estimated odds ratio is 22.39. An increase of one unit in the log concentration of insecticide (x) is associated with a 22-fold increase in the odds that an insect will be killed. The confidence interval for the odds is given in the SAS output: (10.470, 47.896).

Remember that the test of the null hypothesis that the slope is 0 is the same as the test of the null hypothesis that the odds are 1. If we were reporting the results in terms of the odds, we could say, "The odds of killing an insect increase by a factor of 22.3 for each unit increase in the log concentration of insecticide ($X^2 = 64.23$, $P < 0.001$; 95% CI = 10.5 to 47.9)."

17.11 Find the 95% confidence interval for the slope. Using the information in the output of Figure 17.5, find a 95% confidence interval for β_1.

SPSS

Variables in the Equation

	B	S.E.	Wald	df	Sig.	Exp(B)	95.0% C.I. for EXP(B)	
							Lower	Upper
LCONC	3.109	0.388	64.233	1	0.000	22.394	10.470	47.896
Constant	−4.892	0.643	57.961	1	0.000	0.008		

SAS

The LOGISTIC Procedure

Analysis of Maximum Likelihood Estimates

Parameter	DF	Estimate	Standard Error	Wald Chi-Square	Pr > ChiSq
Intercept	1	−4.8923	0.6426	57.9606	< 0.001
lconc	1	3.1088	0.3879	64.2332	< 0.001

Odds Ratio Estimates

Effect	Point Estimate	95% Wald Confidence Limits	
lconc	22.394	10.470	47.896

Minitab

Logistic Regression Table

Predictor	Coef	StDev	Z	P	Odds Ratio	95% CI Lower	95% CI Upper
Constant	-4.8923	0.6426	-7.61	0.000			
lconc	3.1088	0.3879	8.01	0.000	22.39	10.47	47.90

FIGURE 17.5 Logistic regression output from SPSS, SAS, and Minitab for the insecticide data, for Example 17.8.

17.12 **Find the 95% confidence interval for the odds ratio.** Using the estimate b_1 and its standard error, find the 95% confidence interval for the odds ratio and verify that this agrees with the interval given by SAS.

17.13 **X^2 or z.** The Minitab output in Figure 17.5 does not give the value of X^2. The column labeled "Z" provides similar information.

(a) Find the value under the heading "Z" for the predictor lconc. Verify that Z is simply the estimated coefficient divided by its standard error. This is a z statistic that has approximately the standard Normal distribution if the null hypothesis (slope 0) is true.

(b) Show that the square of z is X^2. The two-sided P-value for z is the same as P for X^2.

In Example 17.5 we studied the problem of predicting whether or not the taste of cheese was acceptable using Acetic as the explanatory variable. We now revisit this example to include the results of inference.

Logistic Regression Table

Predictor	Coef	StDev	Z	P	Odds Ratio	95% CI Lower	95% CI Upper
Constant	-13.705	5.932	-2.31	0.021			
acetic	2.249	1.027	2.19	0.029	9.48	1.27	70.96

FIGURE 17.6 Logistic regression output from Minitab for the cheese data with Acetic as the explanatory variable, for Example 17.9.

EXAMPLE 17.9 **Predicting acceptable taste in cheese**

Figure 17.6 gives the output from Minitab for a logistic regression analysis using Acetic as the explanatory variable. The fitted model is

$$\log(\text{ODDS}) = b_0 + b_1x = -13.705 + 2.249x$$

This agrees up to rounding with the result reported in Example 17.5.

From the output we see that because $P = 0.029$, we can reject the null hypothesis that the slope $\beta_1 = 0$. The value of the test statistic is $z = 2.19$, calculated from the estimate $b_1 = 2.249$ and its standard error $\text{SE}_{b_1} = 1.027$. Minitab reports the odds ratio as 9.48, with a 95% confidence interval of (1.27, 70.96).

We estimate that increasing the acetic acid content of the cheese by one unit will increase the odds that the cheese will be acceptable by about 9 times. The data, however, do not give us a very accurate estimate. The odds ratio could be as small as a little more than 1 or as large as 71 with 95% confidence. We have evidence to conclude that cheeses with higher concentrations of acetic acid are more likely to be acceptable, but establishing the true relationship accurately would require more data.

SECTION 17.2 SUMMARY

■ Software fits the data to the model, producing estimates b_0 and b_1 of the parameters β_0 and β_1. Software also produces standard errors for these estimates.

■ A **level C confidence interval for the intercept** β_0 is

$$b_0 \pm z^*\text{SE}_{b_0}$$

A **level C confidence interval for the slope** β_1 is

$$b_1 \pm z^*\text{SE}_{b_1}$$

A **level C confidence interval for the odds ratio** e^{β_1} is obtained by transforming the confidence interval for the slope,

$$\left(e^{b_1 - z^*\text{SE}_{b_1}},\ e^{b_1 + z^*\text{SE}_{b_1}}\right)$$

In these expressions z^* is the standard Normal critical value with area C between $-z^*$ and z^*.

■ The null hypothesis that x does not help predict p in the logistic regression model is H_0: $\beta_1 = 0$ or H_0: $e^{\beta_1} = 1$ in terms of the odds ratio. To test this hypothesis, compute the **test statistic**

$$X^2 = \left(\frac{b_1}{\mathrm{SE}_{b_1}}\right)^2$$

■ In terms of a random variable χ^2 having a χ^2 distribution with 1 degree of freedom, the P-value for a test of H_0 against H_a: $\beta_1 \neq 0$ is approximately $P(\chi^2 \geq X^2)$.

17.3 Multiple Logistic Regression*

The CHEESE data set includes three explanatory variables: Acetic, H2S, and Lactic. Example 17.9 examines the model where Acetic alone is used to predict the odds that the cheese is acceptable. Do the other explanatory variables contain additional information that will give us a better prediction? We use **multiple logistic regression** to answer this question. Generating the computer output is easy, just as it was when we generalized simple linear regression with one explanatory variable to multiple linear regression with more than one explanatory variable in Chapter 11. The statistical concepts are similar, although the computations are more complex. Here is the analysis.

multiple logistic regression

EXAMPLE 17.10 **Multiple logistic regression**

As in Example 17.9, we predict the odds that the cheese is acceptable. The explanatory variables are Acetic, H2S, and Lactic. Figure 17.7 gives the output. From the Minitab output, we see that the fitted model is

$$\begin{aligned}\log(\text{ODDS}) &= b_0 + b_1 \text{ Acetic} + b_2 \text{ H2S} + b_3 \text{ Lactic} \\ &= -14.26 + 0.584 \text{ Acetic} + 0.685 \text{ H2S} + 3.47 \text{ Lactic}\end{aligned}$$

When analyzing data using multiple regression, we first examine the hypothesis that *all* of the regression coefficients for the explanatory variables are zero. We do the same for logistic regression. The hypothesis

$$H_0: \beta_1 = \beta_2 = \beta_3 = 0$$

is tested by a chi-square statistic with 3 degrees of freedom. This is given in the last line of the output. Minitab calls the statistic "G." The value is $G = 16.33$, with

*This optional section uses ideas from Chapter 11.

SPSS

Omnibus Tests of Model Coefficients

	Chi-square	df	Sig.
Model	16.334	3	0.001

Variables in the Equation

	B	S.E.	Wald	df	Sig.	Exp(B)	95.0% C.I. for EXP(B)	
							Lower	Upper
ACETIC	0.584	1.544	0.143	1	0.705	1.794	0.087	37.001
H2S	0.685	0.404	2.873	1	0.090	1.983	0.898	4.379
LACTIC	3.468	2.650	1.713	1	0.191	32.084	0.178	5776.637
Constant	−14.260	8.287	2.961	1	0.085	0.000		

SAS

Testing Global Null Hypothesis: BETA = 0

Test	Chi-Square	DF	Pr > ChiSq
Likelihood Ratio	16.3344	3	0.0010

Analysis of Maximum Likelihood Estimates

Parameter	DF	Estimate	Standard Error	Wald Chi-Square	Pr > ChiSq
Intercept	1	−14.2604	8.2869	2.9613	0.0853
acetic	1	0.5845	1.5442	0.1433	0.7051
h2s	1	0.6848	0.4040	2.8730	0.0901
lactic	1	3.4684	2.6497	1.7135	0.1905

Odds Ratio Estimates

Effect	Point Estimate	95% Wald Confidence Limits	
acetic	1.794	0.087	37.004
h2s	1.983	0.898	4.379
lactic	32.086	0.178	>999.999

Minitab

```
Logistic Regression Table
                                                Odds      95% CI
Predictor       Coef    StDev       Z       P   Ratio   Lower     Upper
Constant     -14.260    8.287   -1.72   0.085
acetic         0.584    1.544    0.38   0.705    1.79    0.09     37.01
h2s           0.6849   0.4040    1.69   0.090    1.98    0.90      4.38
lactic         3.468    2.650    1.31   0.191   32.09    0.18   5777.85

Log-Likelihood = -9.230
Test that all slopes are zero: G = 16.334, DF = 3, P-Value = 0.001
```

FIGURE 17.7 Logistic regression output from SPSS, SAS, and Minitab for the cheese data with Acetic, H2S, and Lactic as the explanatory variables, for Example 17.10.

3 degrees of freedom. The P-value is 0.001. We reject H_0 and conclude that one or more of the explanatory variables can be used to predict the odds that the cheese is acceptable.

Next, examine the coefficients for each variable and the tests that each of these is 0 *in a model that contains the other two*. The P-values are 0.71, 0.09, and 0.19. None of the null hypotheses, H_0: $\beta_1 = 0$, H_0: $\beta_2 = 0$, and H_0: $\beta_3 = 0$, can be rejected. That is, none of the three explanatory variables adds significant predictive ability once the other two are already in the model.

Our initial multiple logistic regression analysis told us that the explanatory variables contain information that is useful for predicting whether or not the cheese is acceptable. Because the explanatory variables are correlated, however, we cannot clearly distinguish which variables or combinations of variables are important. Further analysis of these data using subsets of the three explanatory variables is needed to clarify the situation. We leave this work for the exercises.

SECTION 17.3 SUMMARY

- In **multiple logistic regression** the response variable has two possible values, as in logistic regression, but there can be several explanatory variables.

- As in multiple regression there is an **overall test** for all of the explanatory variables. The null hyphothesis that the coefficients for all of the explanatory variables are zero is tested by a statistic that has a distribution that is approximately χ^2 with degrees of freedom equal to the number of explanatory variables. The P-value is approximately $P(\chi^2 \geq X^2)$.

- Hypotheses about **individual coefficients,** H_0: $\beta_j = 0$ or H_1: $e^{\beta_j} = 1$, in terms of the odds ratio, are tested by a statistic that is approximately χ^2 with 1 degree of freedom. The P-value is approximately $P(\chi^2 \geq X^2)$.

STATISTICS IN SUMMARY

Logistic regression is much like linear regression. We use one or more explanatory variables to predict a response in both instances. For logistic regression, however, the response variable has only two possible values. The statistical inference issues are quite similar. We can test the significance of an explanatory variable and give confidence intervals for its coefficient in the model. When there are several explanatory variables, we can test the null hypothesis that all of their coefficients are zero and the null hypothesis that a single coefficient is zero when the other variables are in the model. Here are the skills you should develop from studying this chapter.

A. PRELIMINARIES

1. Recognize the logistic regression setting: a success-or-failure response variable and a straight-line relationship between the log odds of a success and an explanatory variable x.
2. If the data contain several observations with the same or similar values of the explanatory variable, compute the proportion of successes, the odds of a success, and the log odds for each different value. Plot the log odds versus the explanatory variable. The relationship should be approximately linear.

B. INFERENCE USING SOFTWARE

1. Recognize which type of inference you need in a particular logistic regression setting.
2. Explain in any specific logistic regression setting the meaning of the slope β_1 in the logistic regression model and of the odds ratio e^{β_1}.
3. Understand software output for logistic regression. Find in the output estimates of all parameters in the model and their standard errors.
4. Carry out tests and calculate confidence intervals for the logistic regression slope.
5. Estimate the odds ratio and give a 95% confidence interval for this quantity.

CHAPTER 17 REVIEW EXERCISES

17.14 Holiday purchases. A poll of 811 adults aged 18 or older asked about purchases that they intended to make for the upcoming holiday season.[4] One of the questions asked about what kind of gift they intended to buy for the person on whom they intended to spend the most. Clothing was the first choice of 487 people.

(a) What proportion of adults said that clothing was their first choice?

(b) What are the odds that an adult will say that clothing is his or her first choice?

(c) What proportion of adults said that something other than clothing was their first choice?

(d) What are the odds that an adult will say that something other than clothing is his or her first choice?

(e) How are your answers to parts (a) and (d) related?

17.15 Stock options. Different kinds of companies compensate their key employees in different ways. Established companies may pay higher salaries, while new companies may offer stock options that will be valuable if the company succeeds. Do high-tech companies tend to offer stock options more often than other companies? One study looked at a random sample of 200 companies. Of these, 91 were listed in the *Directory of Public High Technology Corporations*, and 109 were not listed. Treat these two groups as SRSs of high-tech and non-high-tech companies. Seventy-three of the high-tech companies and 75 of the non-high-tech companies offered incentive stock options to key employees.[5]

(a) What proportion of the high-tech companies offer stock options to their key employees? What are the odds?

(b) What proportion of the non-high-tech companies offer stock options to their key employees? What are the odds?

(c) Find the odds ratio using the odds for the high-tech companies in the numerator. Interpret the result in a few sentences.

17.16 Log odds for high-tech and non-high-tech firms. Refer to the previous exercise.

(a) Find the log odds for the high-tech firms. Do the same for the non-high-tech firms.

(b) Define an explanatory variable x to have the value 1 for high-tech firms and 0 for non-high-tech firms. For the logistic model, we set the log odds equal to $\beta_0 + \beta_1 x$. Find the estimates b_0 and b_1 for the parameters β_0 and β_1.

(c) Show that the odds ratio is equal to e^{b_1}.

17.17 Do the inference. Refer to the previous exercise. Software gives 0.3347 for the standard error of b_1.

(a) Find the 95% confidence interval for β_1.

(b) Transform your interval in (a) to a 95% confidence interval for the odds ratio.

(c) What do you conclude?

17.18 Suppose you had twice as many data. Refer to Exercises 17.15 to 17.17. Repeat the calculations assuming that you have twice as many observations with the same proportions. In other words, assume that there are 182 high-tech firms and 218 non-high-tech firms. The numbers of firms offering stock options are 146 for the high-tech group and 150 for the non-high-tech group. The standard error of b_1 for this scenario is 0.2366. Summarize your results, paying particular attention to what remains the same and what is different from what you found in Exercises 17.15 to 17.17.

17.19 Bicycle accidents and alcohol. In the United States approximately 900 people die in bicycle accidents each year. One study examined the records of 1711 bicyclists aged 15 or older who were fatally injured in bicycle accidents between 1987 and 1991 and were tested for alcohol. Of these, 542 tested positive for alcohol (blood alcohol concentration of 0.01% or higher).[6]

(a) What proportion of the bicyclists tested positive for alcohol?

(b) What are the odds that a fatally injured bicyclist will test positive for alcohol?

(c) What proportion of the bicyclists did not test positive for alcohol?

(d) What are the odds that a fatally injured bicyclist will not test positive for alcohol?

(e) How are your answers to parts (a) and (d) related?

17.20 Healthy companies versus failed companies. Case 7.2 (page 476) compared the mean ratio of current assets to current liabilities of 68 healthy firms with the mean ratio for 33 firms that failed. Here we analyze the same data with a logistic regression. The outcome is whether or not the firm is successful, and the explanatory variable is the ratio of current assets to current liabilities. Here is the output from Minitab:

```
Logistic Regression Table
                                                  Odds       95% CI
Predictor        Coef    StDev       Z       P   Ratio   Lower    Upper
Constant      -2.6293   0.6950   -3.78   0.000
ratio          2.6865   0.5632    4.77   0.000   14.68    4.87    44.27
```

(a) Give the fitted equation for the log odds that a firm will be successful.

(b) Describe the results of the significance test for the coefficient of the ratio of current assets to current liabilities.

(c) The odds ratio is the estimated amount that the odds of being successful would increase when the current assets to current liabilities ratio is increased by one unit. Report this odds ratio with the 95% confidence interval.

(d) Write a short summary of this analysis and compare it with the analysis of these data that we performed in Chapter 7. Which approach do you prefer?

17.21 Blood pressure and cardiovascular disease. There is much evidence that high blood pressure is associated with increased risk of death from cardiovascular disease. A major study of this association examined 3338 men with high blood pressure and 2676 men with low blood pressure. During the period of the study, 21 men in the low-blood-pressure and 55 in the high-blood-pressure group died from cardiovascular disease.

(a) Find the proportion of men who died from cardiovascular disease in the high-blood-pressure group. Then calculate the odds.

(b) Do the same for the low-blood-pressure group.

(c) Now calculate the odds ratio with the odds for the high-blood-pressure group in the numerator. Describe the result in words.

17.22 Do the inference and summarize the results. Refer to the previous exercise. Computer output for a logistic regression analysis of these data gives the estimated slope $b_1 = 0.7505$ with standard error $SE_{b_1} = 0.2578$.

(a) Give a 95% confidence interval for the slope.

(b) Calculate the X^2 statistic for testing the null hypothesis that the slope is zero and use Table F to find an approximate P-value.

(c) Write a short summary of the results and conclusions.

17.23 Transform to the odds. The results describing the relationship between blood pressure and cardiovascular disease are given in terms of the change in log odds in the previous exercise.

(a) Transform the slope to the odds and the 95% confidence interval for the slope to a 95% confidence interval for the odds.

(b) Write a conclusion using the odds to describe the results.

17.24 Do syntax textbooks have gender bias? To what extent do syntax textbooks, which analyze the structure of sentences, illustrate gender bias? A study of this question sampled sentences from 10 texts.[7] One part of the study examined the use of the words "girl," "boy," "man," and "woman." We will call the first two words juvenile and the last two adult. Here are data from one of the texts:

Gender	n	X(juvenile)
Female	60	48
Male	132	52

(a) Find the proportion of the female references that are juvenile. Then transform this proportion to odds.

(b) Do the same for the male references.

(c) What is the odds ratio for comparing the female references to the male references? (Put the female odds in the numerator.)

17.25 **Do the inference and summarize the results.** The data from the study of gender bias in syntax textbooks given in the previous exercise are analyzed using logistic regression. The estimated slope is $b_1 = 1.8171$ and its standard error is $\text{SE}_{b_1} = 0.3686$.

(a) Give a 95% confidence interval for the slope.

(b) Calculate the X^2 statistic for testing the null hypothesis that the slope is zero and use Table F to find an approximate P-value.

(c) Write a short summary of the results and conclusions.

17.26 **Transform to the odds.** The gender bias in syntax textbooks is described in the log odds scale in the previous exercise.

(a) Transform the slope to the odds and the 95% confidence interval for the slope to a 95% confidence interval for the odds.

(b) Write a conclusion using the odds to describe the results.

17.27 **Analysis of a reduction in force.** To meet competition or cope with economic slowdowns, corporations sometimes undertake a "reduction in force" (RIF), where substantial numbers of employees are terminated. Federal and various state laws require that employees be treated equally regardless of their age. In particular, employees over the age of 40 years are in a "protected" class, and many allegations of discrimination focus on comparing employees over 40 with their younger coworkers. Here are the data for a recent RIF:

	Over 40	
Terminated	No	Yes
Yes	7	41
No	504	765

(a) Write the logistic regression model for this problem using the log odds of a termination as the response variable and an indicator for over and under 40 years of age as the explanatory variable.

(b) Explain the assumption concerning Binomial distributions in terms of the variables in this exercise. To what extent do you think that these assumptions are reasonable?

(c) Software gives the estimated slope $b_1 = 1.3504$ and its standard error $SE_{b_1} = 0.4130$. Transform the results to the odds scale. Summarize the results and write a short conclusion.

(d) If additional explanatory variables were available, for example, a performance evaluation, how would you use this information to study the RIF?

17.28 **Analysis of the time to repair golf clubs.** The Ping Company makes custom-built golf clubs and competes in the $4 billion golf equipment industry. To improve its business processes, Ping decided to seek ISO 9001 certification.[8] As part of this process, a study of the time it took to repair golf clubs sent to the company by mail determined that 16% of orders were sent back to the customers in 5 days or less. Ping examined the processing of repair orders and made changes. Following the changes, 90% of orders were completed within 5 days. Assume that each of the estimated percents is based on a random sample of 200 orders. Use logistic regression to examine how the odds that an order will be filled in 5 days or less has improved. Write a short report summarizing your results.

17.29 **Know your customers.** To devise effective marketing strategies it is helpful to know the characteristics of your customers. A study compared demographic characteristics of people who use the Internet for travel arrangements and of people who do not.[9] Of 1132 Internet users, 643 had completed college. Among the 852 nonusers, 349 had completed college. Model the log odds of using the Internet to make travel arrangements with an indicator variable for having completed college as the explanatory variable. Summarize your findings.

17.30 **Does income relate to use of the Internet?** The study mentioned in the previous exercise also asked about income. Among Internet users, 493 reported income of less than $50,000 and 378 reported income of $50,000 or more. (Not everyone answered the income question.) The corresponding numbers for nonusers were 477 and 200. Repeat the analysis using an indicator variable for income of $50,000 or more as the explanatory variable. What do you conclude?

17.31 **Gender, alcohol, and fatal bicycle accidents.** A study of alcohol use and deaths due to bicycle accidents collected data on a large number of fatal accidents.[10] For each of these, the individual who died was classified according to whether or not there was a positive test for alcohol and by gender. Here are the data:

Gender	n	X(tested positive)
Female	191	27
Male	1520	515

Use logistic regression to study the question of whether or not gender is related to alcohol use in people who are fatally injured in bicycle accidents.

CHAPTER 17 CASE STUDY EXERCISES

CASE STUDY 17.1: **Predict whether or not the product will be acceptable.** In Examples 17.5, 17.9, and 17.10 we analyzed data from the CHEESE data set described in the Data Appendix. In Examples 17.5 and 17.9, we used Acetic as the explanatory variable. Use each of the other two explanatory variables alone in a logistic regression to predict acceptable taste and summarize the results. With three explanatory variables, there are three different logistic regressions that include two of these. Run these models as well, and summarize the results. Write a report.

CASE STUDY 17.2: **Predict whether or not the GPA will be 3.0 or better.** In Case 11.2 we used multiple regression methods to predict grade point average using the CSDATA data set described in the Data Appendix. The explanatory variables were the two SAT scores and three high school grade variables. Let's define success as earning a GPA of 3.0 or better. So, we define an indicator variable, say HIGPA, to be 1 if the GPA is 3.0 or better and 0 otherwise. Examine logistic regression models for predicting HIGPA using the two SAT scores and three high school grade variables. Summarize all of your results and compare them with what we found using multiple regression to predict the GPA.

CASE STUDY 17.3: **Analyze a Simpson's paradox data set.** In Exercise 2.85 (page 153) we studied an example of Simpson's paradox, *the reversal of the direction of a comparison or an association when data from several groups are combined to form a single group*. The data concerned two hospitals, A and B, and whether or not patients undergoing surgery died or survived. Here are the data for all patients:

	Hospital A	Hospital B
Died	63	16
Survived	2037	784
Total	2100	800

And here are the more detailed data where the patients are categorized as being in good condition or poor condition before surgery:

Good Condition		
	Hospital A	Hospital B
Died	6	8
Survived	594	592
Total	600	600

Poor Condition		
	Hospital A	Hospital B
Died	57	8
Survived	1443	192
Total	1500	200

Use a logistic regression to model the odds of death with hospital as the explanatory variable. Summarize the results of your analysis and give a 95% confidence interval for the odds ratio of Hospital A relative to Hospital B. Then rerun your analysis using the hospital and the condition of the patient as explanatory variables. Summarize the results of your analysis and give a 95% confidence interval for the odds ratio of Hospital A relative to Hospital B. Write a report explaining Simpson's paradox in terms of the results of your analyses.

CASE STUDY 17.4: **Compare the homes in two zip codes.** The HOMES data set described in the Data Appendix gives data on the selling prices and characteristics of homes in several zip codes. In Case 11.3 we used multiple regression models to predict the prices of homes in zip code 47904. For this case study we will look at only the homes in zip codes 47904 and 47906. Define a response variable that has the value 0 for the homes in 47904 and 1 for the homes in 47906. Prepare numerical and graphical summaries that describe the prices and characteristics of homes in these two zip codes. Then explore logistic regression models to predict whether or not a home is in 47906. Summarize your results in a report.

Notes for Chapter 17

1. This example describes an analysis done by Christine Smiley of Kestenbaum Consulting Company, Chicago, Illinois.

2. Logistic regression models for the general case where there are more than two possible values for the response variable have been developed. These are considerably more complicated and are beyond the scope of our present study. For more information on logistic regression, see A. Agresti, *An Introduction to Categorical Data Analysis*, Wiley, 1996; and D. W. Hosmer and S. Lemeshow, *Applied Logistic Regression*, Wiley, 1989.

3. This example is taken from a classical text written by a contemporary of R. A. Fisher, the person who developed many of the fundamental ideas of statistical inference that we use today. The reference is D. J. Finney, *Probit Analysis*, Cambridge University Press, 1947. Although not included in the analysis, it is important to note that the experiment included a control group that received no insecticide. No aphids died in this group. We have chosen to call the response "dead." In the text the category is described as "apparently dead, moribund, or so badly affected as to be unable to walk more than a few steps." This is an early example of the need to make careful judgments when defining variables to be used in a statistical analysis. An insect that is "unable to walk more than a few steps" is unlikely to eat very much of a chrysanthemum plant!

4. The poll is part of the American Express Retail Index Project and is reported in *Stores*, December 2000, pp. 38–40.

5. Based on Greg Clinch, "Employee compensation and firms' research and development activity," *Journal of Accounting Research*, 29 (1991), pp. 59–78.

6. Data from Guohua Li and Susan P. Baker, "Alcohol in fatally injured bicyclists," *Accident Analysis and Prevention*, 26 (1994), pp. 543–548.

7. From Monica Macaulay and Colleen Brice, "Don't touch my projectile: gender bias and stereotyping in syntactic examples," *Language*, 73, no. 4 (1997), pp. 798–825.

8. Based on Robert T. Driescher, "A quality swing with Ping," *Quality Progress*, August 2001, pp. 37–41.

9. From Karin Weber and Weley S. Roehl, "Profiling people searching for and purchasing travel products on the World Wide Web," *Journal of Travel Research*, 37 (1999), pp. 291–298.

10. See Note 6.

Chapter 17

17.1 Exclusive territories: $\hat{p} = 0.761$, ODDS = 3.176. No exclusive territories: $\hat{p} = 0.536$, ODDS = 1.154.

17.3 Exclusive territories: log(ODDS) = 1.156. No exclusive territories: log(ODDS) = 0.143.

17.5 $b_0 = 0.143$; $b_1 = 1.013$; $\text{ODDS}_{\text{exclusive}}/\text{ODDS}_{\text{no exclusive}} = 2.754$, so that the odds for exclusive territories are 2.754 times the odds for no exclusive territories.

17.7 $\text{ODDS}_{x+1}/\text{ODDS}_x = (e^{-13.71} \times e^{2.25(x+1)})/(e^{-13.71} \times e^{2.25x}) = 9.49$.

17.9 Using software, the test of H_0: odds ratio = 1 has a P-value of 0.018, so there is good evidence that the odds for the two groups are different. The 95% confidence interval for the odds ratio is (1.19, 6.36).

17.11 (2.3485, 3.8691).

17.13 (a) $Z = b_1/\text{SE}_{b_1} = 8.014$. (b) $Z^2 = 8.014^2 = 64.224$, the value of the chi-square statistic.

17.15 (a) High-tech: $\hat{p} = 0.802$, ODDS = 4.056. (b) Non-high-tech: $\hat{p} = 0.688$, ODDS = 2.206. (c) $\text{ODDS}_{\text{high-tech}}/\text{ODDS}_{\text{non-high-tech}} = 1.839$, so that the odds of offering stock options for high-tech companies are 1.839 times the odds for non-high-tech companies.

17.17 (a) (−0.047, 1.265). (b) (0.954, 3.543). (c) Since the value 1 is included in the 95% confidence interval for the odds ratio, there is no evidence at the 5% level of significance that the odds of offering stock options differ for high-tech and non-high-tech companies.

17.19 (a) $\hat{p} = 0.317$. (b) ODDS = 0.4636. (c) $\hat{p} = 0.683$. (d) ODDS = 2.1568. (e) The answer in (c) is 1 minus the answer in (a), and the answer in (d) is 1 divided by the answer in (b), except for round-off error.

17.21 (a) $\hat{p} = 0.0165$, ODDS = 0.0168. (b) $\hat{p} = 0.0078$, ODDS = 0.0079. (c) $\text{ODDS}_{\text{high-blood-pressure}}/\text{ODDS}_{\text{low-blood-pressure}} = 2.127$, so that the odds of dying from cardiovascular disease for the high-blood-pressure group are 2.127 times the odds for the low-blood-pressure group.

17.23 (a) The estimate of the odds ratio is $e^{b_1} = 2.118$. A 95% confidence interval for the odds ratio is (1.278, 3.511). (b) The odds of dying from cardiovascular disease for the high-blood-pressure group differ from the odds for the low-blood-pressure group at the 5% level of significance since the value 1 is not included in the confidence interval.

17.25 (a) (1.0946, 2.5396). (b) $\chi^2 = (b_1/\text{SE}_{b_1})^2 = 4.930$. (c) There is good evidence that the slope is not equal to zero, which supports the alternative hypothesis that there is gender bias in syntax textbooks.

17.27 (**a**) $\log[p_i/(1 - p_i)] = \beta_0 + \beta_1 x_i$. (**b**) The probability of being terminated should be constant within each population (over and under 40), and the termination of individuals should be independent. If we are working within a single company or group of similar companies, these assumptions should be approximately satisfied. If the workers are from a variety of industries or a heterogeneous population, then the assumption of constant probability of termination within each group may be violated. (**c**) The estimate of the odds ratio is 3.859, and a 95% confidence interval for the odds ratio is (1.718, 8.670). Since the interval doesn't include 1, there is evidence at the 5% level that the odds of being terminated are different for the two groups. (**d**) If additional explanatory variables were available, they could be included in a multiple logistic regression model.

17.29 The test of H_0: odds ratio = 1 has a *P*-value of 0.000, so there is very strong evidence that the odds of having completed college are different for the two groups. The estimated odds ratio is 1.90, so the odds of having completed college for those using the Internet to make travel arrangements are estimated to be 1.90 times the odds for those who do not use the Internet to make travel arrangements.

17.31 The test of H_0: odds ratio = 1 has a *P*-value of 0.000, so among those having fatal bicycle accidents there is very strong evidence that the odds of having a positive test for alcohol are different for males and females. The estimated odds ratio is 3.11, so the odds of testing positive for alcohol for males are estimated to be 3.11 times the odds for females.